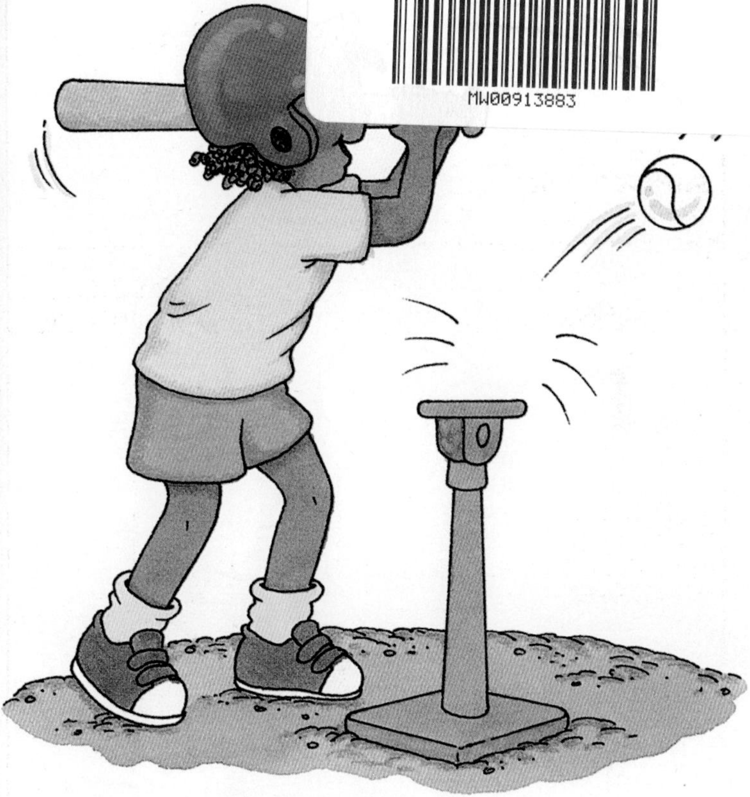

Snap!

Dot zaps it!

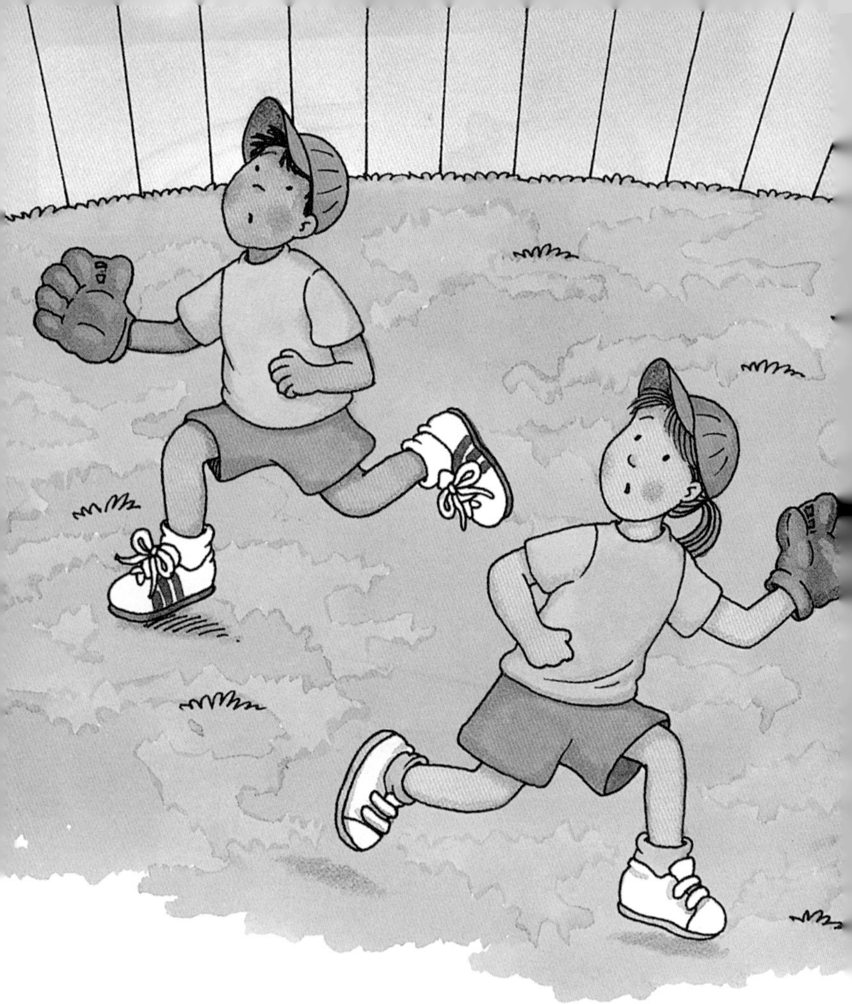

Stan and I split.

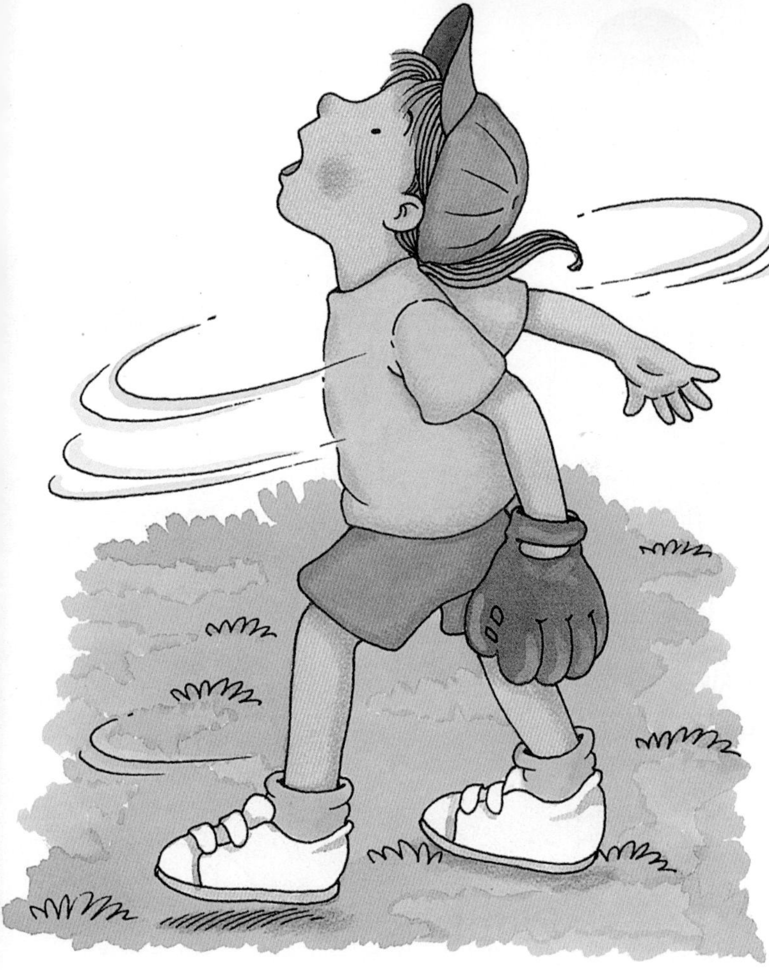

I spin and spin.

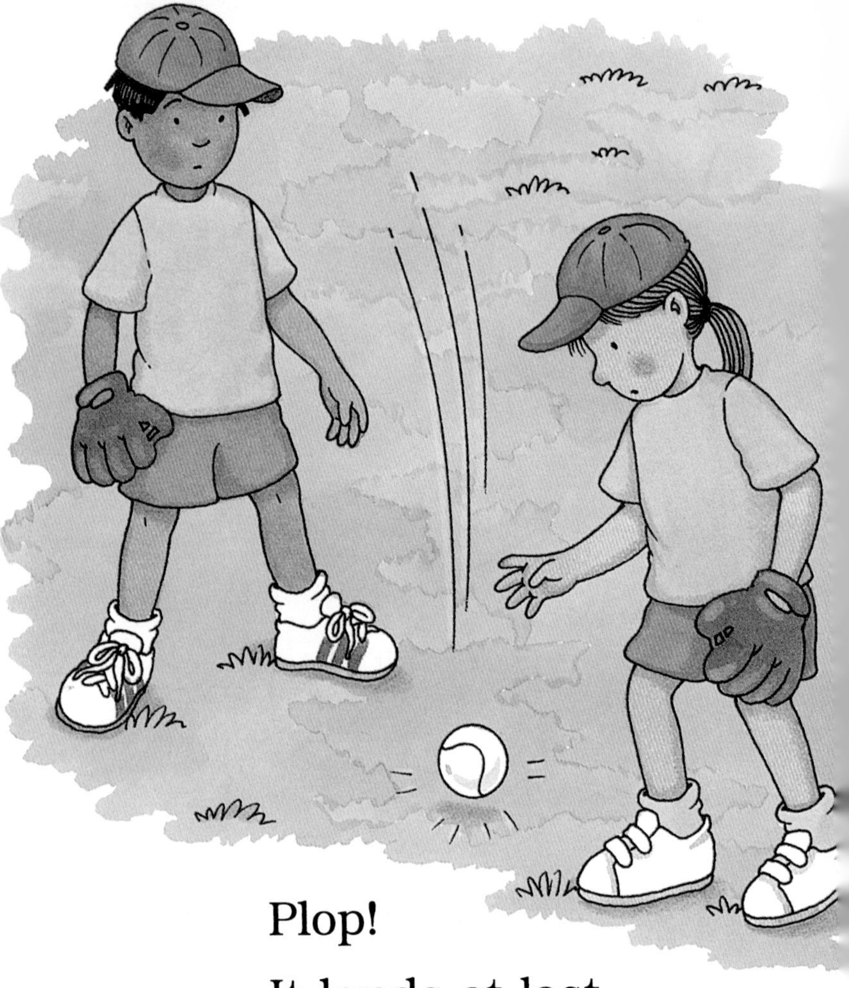

Plop!

It lands at last.

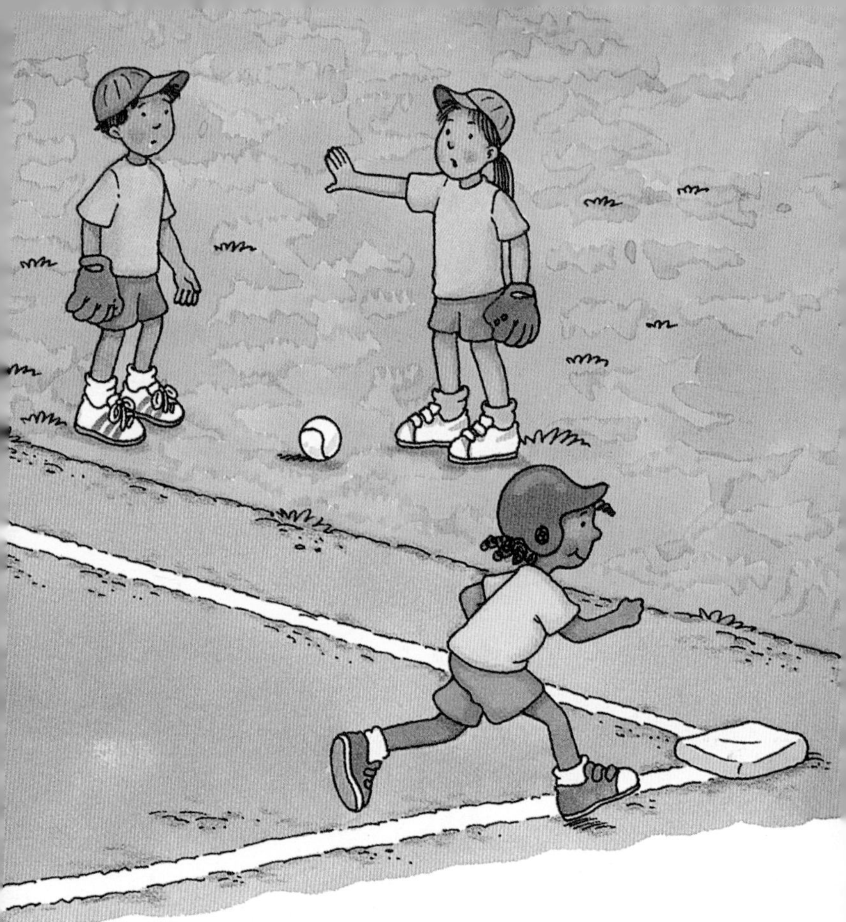

I said, "No!"

Stan stands still.

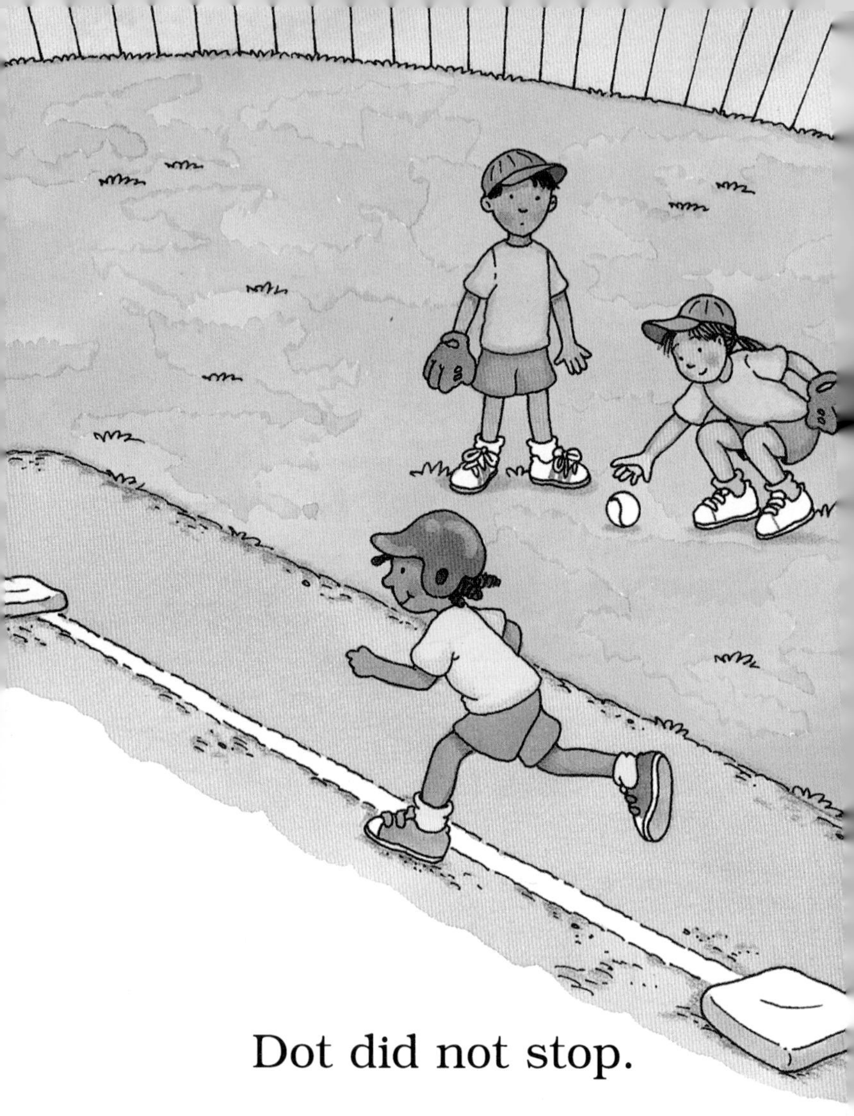

Dot did not stop.

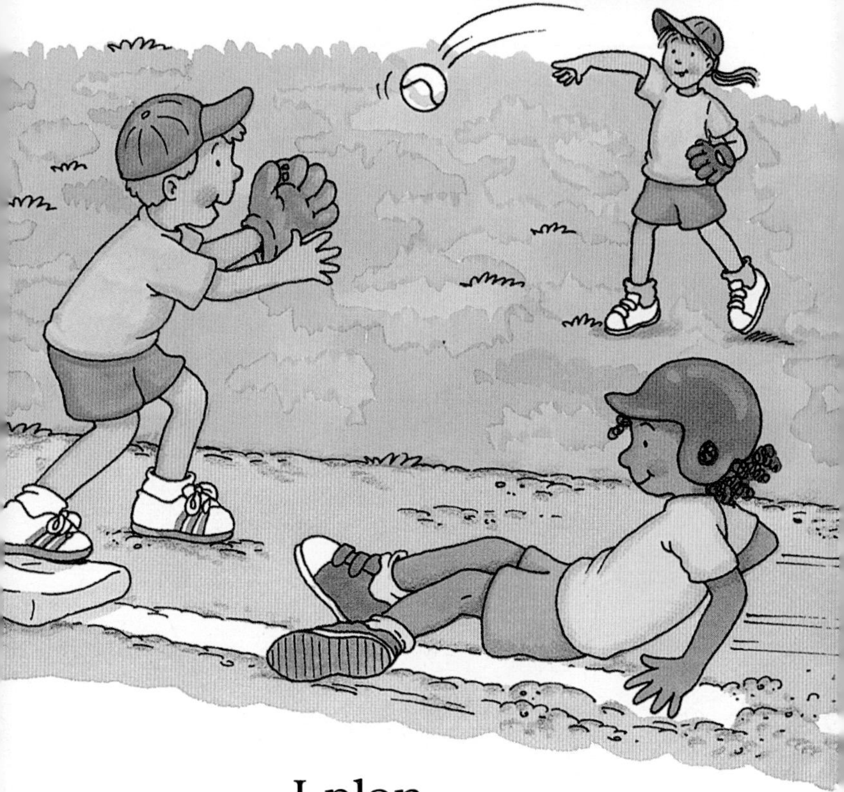

I plan.

I toss it.

Dot slid.

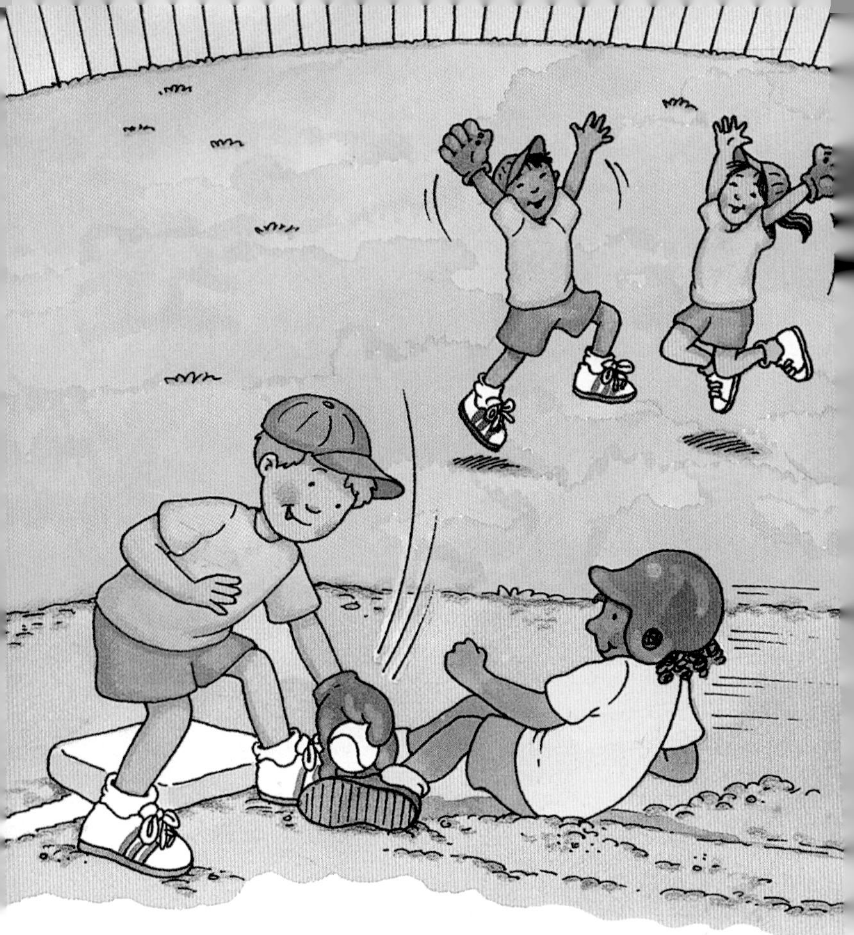

Pat taps Dot.

The End